Crewed Spacecraft

Patrick H. Stakem

(c) 2017, 2022

1st edition

Number 8 in the Space Series

Table of Contents

Table of Contents..2
Introduction...4
A note on Units..4
Author..5
What is space travel, and where does space start?..................7
    Von Braun's and Ley's vision – the Mars Mission..............8
Precursor missions..8
United States..9
    Winged craft into space..9
        X-15...10
        Dynasoar..11
        STS-Shuttle...12
    Space Capsules..17
        Mercury 17
        Gemini..19
        Apollo...20
        Ares..27
        Orion..27
    Space Stations...29
        Manned Orbiting Lab 29
        Skylab...31
    Skylab-II..33
    Shuttle-Mir Program..34
        China..35
        International Space Station – ISS 35
NASA's Commercial Crew Development Program (CCDev).......38
Soviet Union/Russia..39
    Vostok...39
    Voskhod..40
    Soyuz 41
    Buran, the Russian Shuttle...44
Joint Missions...44
    Apollo-Soyuz – the match-up......................................45
Chinese Missions..46

    Shenzhou..46
ESA..47
ISRO..47
Private efforts..48
    Blue Origin..48
    Virgin Galactic...50
    SpaceX...51
    Sierra Nevada's Dreamchaser..53
    Boeing CST-100 Starliner...54
    Excalibur Almaz..55
    Stratolaunch Systems..55
Space Tourists 55
The fallen Astronauts memorials..57
Afterword..57
Bibliography...58
    Resources...72
Glossary..74
If you enjoyed this book, you might also be interested in some of these...80

"The only way to discover the limits of the possible is to go beyond them into the impossible." Sir Arthur C. Clarke

Introduction

Crewed spacecraft means spacecraft with humans aboard. Young, or old, male or female, space travel is an equal opportunity hazardous endeavor. As of this writing, three spacefaring nations have put 559 humans into space, the Soviet Union/Russia, the United States, and China. Numerous astronauts from many other nations have hitched a ride. We are going to discuss historical, ongoing, and future efforts of getting people to space. We will mention in passing Space Stations in orbit, but that is a big topic, and well best be covered in a companion volume. All material was derived from non-ITAR sources. Currently, besides the United States, there are 25 countrys with launch vehicles. This work is focused on the United States. There are three space stations, the ISS, Chinese, and the pending India one.

A note on Units

I am fairly conversant in both English and Metric units (what is the metric equivalent of furlongs per fortnight?). Metric (SI) is mandated for NASA usage now, for interchangeability with our partner space faring nations. When a lot of the legacy flights discussed here were flown, English units were the norm. I have tried to keep the units comparable to the mission at the time. Conversions are easy enough, but units conversion is a source of error. It's not what you know about units and measurement, its how you think. And, I still think English units (even the English use Metric now), and convert in my head or on my phone.

For scientific/engineering work, the Metric system is well thought out. For artisans, the English system served well, as most units were divided by 2. Which is easy. Fold the cloth. Hopefully, when we are all taught Metric first, some one will still remember the conversions. You just need a slide rule....

Author

Mr. Patrick H. Stakem has been fascinated by the space program since the Vanguard launches in 1957. He received a Bachelors degree in Electrical Engineering from Carnegie-Mellon University, and Masters Degrees in Physics and Computer Science from the Johns Hopkins University. At Carnegie, he worked with a group of undergraduate students to re-assemble, modify, and operate a surplus missile guidance computer, which was later donated to the Smithsonian. He was brought up in the mainframe era, and was taught to never trust a computer you could lift.

He began his career in Aerospace with Fairchild Industries on the ATS-6 (Applications Technology Satellite-6) program, a communication satellite that developed much of the technology for the TDRSS (Tracking and Data Relay Satellite System). He followed the ATS-6 Program through its operational phase, and worked on other projects at NASA's Goddard Space Flight Center including the Hubble Space Telescope, the

International Ultraviolet Explorer (IUE), the Solar Maximum Mission (SMM), some of the Landsat missions, and Shuttle. He was posted to NASA's Jet Propulsion Laboratory for Mars-Jupiter-Saturn (MJS-77), which later became the *Voyager* mission, and which is still operating and returning data from outside the solar system at this writing. He initiated and lead the international Flight Linux Project for NASA's Earth Sciences Technology Office. He is the recipient of the Shuttle Program Manager's Commendation Award, and has completed 42 NASA Certification courses. He has two NASA Group Achievement Awards, and the Apollo-Soyuz Test Program Award.

Mr. Stakem was affiliated with the Whiting School of Engineering of the Johns Hopkins University, the Graduate Computer Science Department of Loyola University in Maryland, and Capital Technology Institute. He supported the Summer Engineering Bootcamp Projects at Goddard Space Flight Center for 2 years. These resulted in the Greenland Rover, a tracked robot measuring the depth of the ice sheet.

Mr. Patrick H. Stakem has been fascinated by the space program since the Vanguard launches in 1957. He received a Bachelors degree in Electrical Engineering from Carnegie-Mellon University, and Masters Degrees in Physics and Computer Science from the Johns Hopkins University. At Carnegie, he worked with a group of undergraduate students to re-assemble, modify, and operate a surplus missile guidance computer, which was later donated to the Smithsonian. He was brought up in the mainframe era, and was taught to never trust a computer you could lift.

He supported the Summer Engineering Bootcamp Projects at Goddard Space Flight Center for 2 years. These resulted in the Greenland Rover, a tracked robot measuring the depth of the ice sheet.

Now he supports International collaborative open-source projects, and STEM Programs.

Mr. Stakem can be found on Facebook and LinkedIn. Comments, corrections, suggestions are appreciated.

What is space travel, and where does space start?

The *Fédération Aéronautique Internationale* defines space as beginning at 100 km altitude. Two of the 199 X-15 flights managed to reach that altitude, so the pilots were officially astronauts. So far, 3 nations have launched some of their citizens into space, with the former Soviet Union being the first. The United States followed along, and, more recently, China has sent some of their Citizens to orbit. There have been space-farers from over 40 countries, taken along on shared missions by the craft of the major space-faring nations, China, Russia, and the U. S. The International Space Station is truly an International effort.

Von Braun's and Ley's vision – the Mars Mission

Wernher von Braun produced a detailed technical project plan for getting to Mars in 1952, but he had been thinking about it long before that. He and fellow rocket enthusiast Willy Ley worked on Mars Mission Concepts, a long time before the first person rode a rocket.

Precursor missions

Before launching humans into space, there were suborbital launches, and even earlier, a pilot officially reached space (100km altitude) in a rocket plane, the X-15. Capsule flights with dogs (Soviet Union) and chimpanzees (U. S.) preceded human attempts. Astro-chimp Ham was the first hominid launched into space, in 1961. Ham had been born in Cameroon, Africa. He emigrated to the U.S. And was purchased by the U. S. Air Force from a rare animal farm in Florida.

At Holloman Air Force Base, he received his training in completing simple, timed tasks. He was not a passenger, but proved that primates could function in space, after the harsh launch environment. His January, 1961 flight prepared the way for an advanced hominid, Alan Sheppard, to conduct a sub-orbital flight in May of 1961. Sheppard, however, did not receive banana pellets for a job well done.

Ham wore a fully functional space suit, and that may have saved him during a partial loss of pressure in the capsule. His flight duration was just above 16 ½ minutes, from take-off to splashdown in the Atlantic. Ham retired to the National Zoo in Washington, D.C. for 17 years after his groundbreaking flight. He then transferred to a chimp enclosure in North Carolina, where he died in 1983. His remains are buried at the International Space Hall of Fame in Alamogordo, NM. Godspeed, Ham!

The Russians used dogs in their early flights. The requirements were that they could weigh no more than 15 pounds, be no more than 14 inches long, have a calm temperament, and be photogenic. Supposedly, these were strays from the streets of Moscow. Laika, in 1957, became the first creature from Earth to enter space. Unfortunately for her, it was a one-way journey. Two additional dogs following in her paw prints, made it back successfully. These were named Belka and Strelka.

United States

This sections discusses the crewed space missions by the United States. This will cover the various capsules and space planes. In depth coverage of Space Stations and Rocket Planes will be the subject of their own books in this series.

Winged craft into space

This section discusses winged rocket planes. Jet engines can't handle the altitude since they run out of air for combustion. The rocket carries its own oxidizer. There were all liquid fuel assemblies. A lot of rocket plane development was done in Germany before and during world War-II. This is explored in more detail in another of the author's book.

X-15

The X-15 was a crewed hypersonic aircraft, designed to fly to the edge of space, defined as 100 km . A Pilot who reached this altitude was officially an Astronaut. The X-15 was carried to altitude under the wing on a B-52 Mothership. After it was released, its rocket engine was ignited, pushing it up out of most of the atmosphere. There were no air breathing engines. It did a "dead-stick" landing. There were two flight to the 100 km altitude, both by Joseph A. Walker in 1963. The X-15 flights were USAF, U.S. Navy, and NASA sponsored. The first flight was in 1959, and flights continued for 9 years.

The X-15 holds the official world's record to the highest speed recorded by a crewed powered aircraft 4,250 mph (Mach 6.7) at 102,100 feet altitude. That was in 1967. It is officially the world's first space plane. It set the World record for the highest speed in a powered aircraft in 1967 at 4,250 miles per hour, and that record stands at this writing.

The X-15 was developed from a concept of Dr. Walter Dornberger for NACA, predecessor of NASA, in 1954. He was one of the captured German Rocket Team members that formed the core of the U.S. Space effort. Dornberger served as the military officer in charge of the German rocket program, and he came to the U.S. with von Braun.

The X-15 operated in two distinct domains, and had both aerodynamic control surfaces, and rocket thrusters. The plane included a pilot ejection seat, usable up to Mach 4, and 120,000 feet. The main engines were dual *Reaction*

Motors XLR11 units, using alcohol and liquid oxygen to achieve a total of 16,000 pounds of thrust. Earlier, a single XLR11 pushed the Bell X-1 to be the first aircraft to exceed the speed of sound (Mach 1). The X-15 was later fitted with the upgraded XLR99 for 57,000 pounds of thrust. Over 175 flights were made in that configuration. The fleet of three X-15's made a total of 199 test flights, the last in 1968. Twelve pilots flew the planes, including future astronaut Neil Armstrong, who would go on to become the first man on the moon.

The pilot had controls both for the aerodynamic surfaces, and the reaction control thrusters. A later improvement had the Honeywell MH-96 Flight computer use the appropriate system, with pilot input. This was an early fly-by-wire system, which is now common in almost all aircraft.

X-15's can be seen today at the National Air and Space Museum in Washington, D.C. and the National Museum of the United States Air Force, located at Wright-Patterson Air force Base in Dayton, Ohio.

Dynasoar

Dynasoar was a USAF Project for a winged spacecraft. It was developed by Boeing as the X-20. The project started in 1957, and was canceled in 1963, just at the beginning of construction. The trend in spacecraft went to space capsules that had heat shields, and returned on a ballistic trajectory to a ground or water landing. Dynasoar was designed to reach Earth orbit with a single pilot, using a Titan launch vehicle. It had an equipment bay behind the pilot, that could be used for payloads. A variant, the X-20X had a rear crew compartment that could hold 4. It was reusable, like the Shuttle would be. Both included a trans-stage at the rear for orbital maneuvering, that would be jettisoned before reentry. The plane could dip into the atmosphere and back to change its orbital inclination, without a large expenditure of fuel, which an orbiting spacecraft would need,

The project went back to Dr. Dornberger, from the German World War-II rocketry efforts. He had detailed knowledge of Eugene Sanger's Silbervogel Spaceplane project. The Dynasoar was modeled on that. Quite a few studies were done by the major U. S. Aerospace companies. Dynasoar was to be a successor to the X-15 research vehicle. The contract to build the vehicle was awarded to Boeing. Later in the program, seven astronauts were chosen from NASA and the Air Force to fly the Dynasoar, including Neil Armstrong.

In retrospect, the program was canceled due to uncertainty over the booster to be used, and a lack of planning and clear goals. The program kept changing requirements, and thus no one agreed on what to build. The Dynasoar did influence the later Space Shuttle design and operations.

STS-Shuttle

The Space Transportation System (STS) was a crewed launch and recovery system for spacecraft, that used rocket propulsion to achieve orbit, and glided back to Earth to land on a run-way. A major advantage of the

Shuttle system was, when it carried a spacecraft to orbit, it could check to see if it survived the harsh launch environment. If not, the Shuttle could bring it home. Perhaps its major achievement was to repair the Hubble Space Telescope in orbit over several missions. The Shuttles made several repair trips to the Hubble Space Telescope, to work around its optical problem, and change out some failed computers and such. The Shuttle was also instrumental in assembling the International Space Station in orbit.

At launch, the STS consisted of the winged Shuttle vehicle, a large liquid fuel and oxidizer external tank, and two solid rocket boosters. The solid rocket casings were retrieved from the ocean, and refurbished and reused. The external tanks were not recovered, and were targeted away from shipping lanes in the Pacific and Indian oceans.

There was a mock-up, a prototype, and five flight units. Two of the flight units were destroyed, one at launch, one at reentry, with loss of crew. The Shuttle could accommodate up to 8 crew. There were 135 flights. The Shuttle returned to a runway landing at Kennedy Space Center, or Edwards Air Force Base.

The Shuttle Orbiter rode the side of a large fuel and oxidizer tank to orbit, assisted by fall-away solid boosters to get everything going. The boosters fell into the ocean after they finished their burn, and were recovered and reused. The Shuttle Orbiter had three engines, fed from the large external tank. When the engines had burned sufficiently to achieve orbit, the Orbiter separated from the tank, which fell into the ocean, but was not recovered. The Orbiter continued to its destination altitude. The engines went with it, but no longer had a source of fuel or oxidizer. He Orbiter could adjust its orbit somewhat with its OMS (orbital maneuvering system) engines, using fuel onboard. There were also (reaction control system) RCS engines to adjust attitude. Upon reentry, the Shuttle flew in a nose-up attitude, as the bottom of the craft and wings were covered in heat-resistant tiles. After sufficient atmosphere was reached, the aircraft control surfaces could be used, and the Orbiter was flown like a plane to a runway landing. Well, like a 165,000 lb glider. No air-breathing engines were included, although this option was discussed early in the program.

Ideally, the Orbiter landed at the runway back at the Launch site, and could be easily towed to the maintenance facility. Another option was to use the Dryden flight facilities vast expanses of hard desert. In that case, the Shuttle was brought back to the Kennedy Space Center on the back of a specially modified 747 transport aircraft.

There was a plan to launch Shuttles from Vandenberg Air Force Base in California, which would allow them to go to polar orbit. This was not implemented. There was also a two-stage rocket called the Interim Upper Stage or Inertial Upper Stage (IUS) which would deploy from the Shuttle bay with a payload going to a higher orbit. It could also be used with a Titan booster. One IUS mission put the Ulysses Mission to study the polar regions of the Sun. Quite a few were used to put the Tracking and Data Relay Satellites (TDRS) into higher orbit.

A versatile craft, the Shuttle could take satellites to orbit, check them out, release them if they worked fine or bring them back, if they didn't.

NASA re-purposed the Apollo-era Vertical Assembly building (VAB) at the Cape to assemble the Shuttle stack. The solid boosters were bolted down on the crawler/transporter base, and the large external tank and Shuttle Orbiter were hoisted up and attached. The building is so large, it has its own weather inside. There are two launch pads, 39-A and 39-B, essentially identical.

The launch sequence proceeded according to a well defined procedure. The three liquid engines were ignited one at a time in sequence, to check that they were all working properly. This pushed the orbiter's nose forward about a meter. When the liquid engine performance was verified, the explosive bolts holding the solid boosters to the pad were blown, and the solid boosters were ignited. Then, you were on your way. At this point the orbiter's nose rebounded, a movement the astronauts call "the twang."

In some videos of a Shuttle launch, you will see a series of sparks below the engines. That was to ignite any residual or leaked hydrogen from the external tank. A huge water spray was started before engine ignition. This was to partially protect the pad, but also to damp the acoustic energy from the engines. Otherwise it reflected up onto the vehicle, and could do damage.

The Space Shuttles carried five identical computers, the circa-1972 AP-101's, derived from the IBM System/360 System/4 Pi mainframe architecture. It was a 32-bit machine with 16 registers. It had an instruction set of 154 opcodes. One of the five AP-101's on the Shuttle contained software derived independently from the software loaded on the other four. Each unit had a CPU and an IOP - Input/Output Processor. Each IOP had 24 channels, each with its own bus and processor. Triple redundant power supplies, fed by separate essential electrical buses were used. The computers were located in three separate locations in the Shuttle Orbiter. Redundancy is everything.

Want to see Shuttle flight hardware? *Pathfinder*, a full-size mock-up, is at the Alabama Space and Rocket Center, Huntsville, AL. OV-101 *Enterprise*, a prototype used for flight tests in the atmosphere, is at the Intrepid Sea, Air & Space Museum in New York City. OV-102 *Columbia* was destroyed (with loss of crew) in a re-entry accident on February 1, 2003. OV-099 *Challenger* was *d*estroyed (with loss of crew) in a launch accident, January 28, 1986. Debris was recovered and is stored, sealed in an old missile silo, at Cape Canaveral Air Station, FL. OV-103 *Discovery* rests in the National Air and Space Museum, Steven F. Udvar-Hazy Center, Chantilly, VA. (near Dulles Airport). OV-104 *Atlantis* may be seen at the Kennedy Space Center, Cape Canaveral, FL. OV-105 Endeavor is at the Samuel Oschin Pavilion of the California Science Center in Los Angeles, CA.

No flown external tanks have survived, but unused ET-94 is in Los Angeles and will be on display with Space Shuttle Endeavor at the California Science Center, when the Samuel Oschin Air and Space Center opens in 2018. Three external tanks were in manufacturing when the Shuttle Program ended, numbers ET-139-141.

One of the Shuttle Carrier aircraft, a specially adapted Boeing 747, can be seen at Palmdale (CA)'s Joe Davis Heritage Airpark. NASA retains ownership of the aircraft. The other carrier aircraft was placed at Space Center Houston, with a Shuttle mock-up on top.

Space Capsules

This section discusses various space capsules, designed to re-enter atmosphere ballisticly, using a heat shield. There's generally no maneuvering during reentry. These capsules can do a water or ground landing.

Mercury

Project Mercury was focused on getting an American into space. The hardware included a single-person capsule from McDonnell aircraft, and a rocket adapted from an ICBM. The Mercury capsule carried a single person, with a limited duration on orbit. It incorporated a launch escape system, to separate the capsule from a malfunctioning rocket. Several suborbital flights were made with chimpanzees and, later, Astronauts, before the first orbital flight. Orbital Mission duration was 3 orbits, about 4 1/2 hours. The demonstrated longest flight time was 34 hours. There were 6 flights of the Mercury spacecraft, 2 of these were sub-orbital

Alan Shepard was the first person to ride in one, a suborbital flight in May of 1961. The sub-orbital flights used the Redstone launch vehicle, but a larger Atlas rocket was used to achieve orbit. John Glenn, the first American to orbit the Earth, rode one in 1962.

The early Mercury flights were ballistic, with no control from the capsule. The Mercury capsule had no onboard computer. All calculations were done on IBM 701 mainframes on the ground, and radioed to the astronaut. The critical parameter was when to fire the retro rockets for reentry.

The capsule used retrorockets to slow down and begin the reentry process. The Astronaut rode down "Backwards", looking opposite to the direction of flight, with a large ablative heat shield to protect the capsule from the heat of reentry. Essentially, the energy used to get the craft to orbit was dissipated as heat when it came back. Like driving up a mountain, and using the brakes to come down safely. At a proper altitude, a parachute was deployed, and the capsule did a water landing. The astronaut and the capsule were recovered by Navy helicopter.

The capsule was tracked around the globe. This required NASA to have a global network of tracking stations, ships, and aircraft, and a control center to monitor the flight. As an aside, one of my old bosses at NASA was in charge of setting up the tracking ships.

Gemini

The Gemini was a two-person capsule, launched to orbit on an Titan-II vehicle from launch complex 19 at Cape Canaveral Air Station, FL (Now, Kennedy Space Center). It was a follow-on to the earlier Mercury Project, and proved designs and concepts for the following Apollo Project. The program ran from 1961 to 1966, with test flights in 1965 and 66, 10 flights in all. Gemini proved, in orbit, missions as long in duration as would be necessary for the lunar flights were feasible. Also, Gemini proved the feasibility of extra-vehicular activity, with space-suited astronauts outside the capsule, worked fine. Gemini also helped validate the procedures for rendezvous and docking of two spacecraft, essential for Apollo. There were two un-crewed flights in 1964-1965 to test the heat shield.

An IBM guidance computer was used on all Gemini flights, including the first rendezvous mission, with Gemini 6 (Astronauts Walter Schirra and Thomas Stafford) and Gemini 7 (Astronauts Frank Borman and James Lovell). Occupying only 1.35 cubic feet of space, the onboard system performed some 7,000 calculations a second to bring the two vehicles nose-to-nose, 120-feet apart, 185 miles above Hawaii. The computer had a memory system capable of holding nearly 20,000 bytes of information.

The Air Force eyed using the Gemini capsule for their own Manned Orbiting Lab (MOL) project, but that program was canceled before any flights were made.

The Gemini-TTV (towed test vehicle) used a large paraglider and deployable wheels to enable the capsule to land at an airfield. Testing was done at Edwards Air Force Base in California. This program was canceled in 1964. After de-orbit, the regular Gemini capsules landed in the ocean, as had Mercury. Mission duration's were 3 orbits, about 4-5 hours.

Apollo

Apollo was a three-person spacecraft, designed for the lunar landings. The Apollo spacecraft were the payloads on the Saturn Ib and V vehicles. These included the Command Module that held the 3 astronauts, the Service module that contained equipment, and the lunar module. The Lunar lander bases were left on the lunar surface. After the crew returned to the Command Module, the ascent stage of the lander was crashed into the lunar surface, to provide controlled moon-quakes for seismometers on the surface. The Command Modules were jettisoned before reentry into Earth's atmosphere.

The Apollo spacecraft was used in 15 flights, 9 of which were lunar missions, with 6 lunar landings.

The early Apollo flights to Earth orbit used the Saturn-I vehicle. The lunar flights used the larger Saturn-V. These were built specially for that purpose.

The Saturn vehicles were developed by the von Braun team at Marshall Space Flight Center, formally the Army's Redstone Arsenal, in Huntsville, Alabama. Von Braun and his team of scientists and engineers had been brought to the U.S. by the Army after World War II ended. The rocket program was kicked off during the early post-World War-II Cold War era by President Eisenhower. At the time, the United States was in a race to space, and particularly, a launch vehicle race, with the Soviet Union. Each U. S. military service, the Army, Navy, and Air Force were developing their own rockets. Inter-service rivalry was finally squashed by Secretary of Defense Charles Wilson, who decided in November of 1956 to make the Air Force the primary missile developer for long range ballistic and space missions. The specifications for a heavy-lift vehicle were developed by the Advanced Research Projects Agency (ARPA).

The Apollo payload consisted of the Launch Escape System, the Apollo capsule, the service module, and the lunar lander. The launch escape system (LES) was located above the Apollo capsule and was jettisoned early in flight. The Lunar Excursion Module (LEM) was stored behind the service module. Once in Earth orbit, the capsule and Service Module were separated, the capsule rotated 180 degrees, and docked to the Lunar module. The lunar package was then separated from the third stage. The capsule, lander, and service module left Earth orbit heading for the moon, while the Third stage was commanded into a solar orbit, to get it out of the way.

The Command Module, or Apollo capsule, was the cockpit and living quarters for the three astronauts. The computing heart of the capsule was the unique Apollo Guidance Computer. The need for a computer onboard the Apollo was required by the chosen approach to the mission, lunar-orbit rendezvous. Part of the spacecraft (Command Module) would remain in lunar orbit, while a detachable part (LEM) would descend to the surface. Later, the LEM would return to lunar orbit and rendezvous with the Command Module, which would then leave lunar orbit and return to Earth. The ability of the Command Module and LEM to do in-flight computations was crucial to this approach. At the time, the only guidance computes were developed for ballistic missiles.

The Service Module was located behind the Command Module, and the astronauts had no direct access to it. It was unpressurized, and contained a re-startable liquid rocket engine and associated propellant, fuel cells, and electronics to support the mission. The fuel cells used hydrogen and oxygen, and some oxygen was also used to replenish the Command Module atmosphere. It had a reaction control system to adjust the spacecraft attitude. The service module had radiators to dump excess heat, and a high gain antenna to communicate with Earth. The Command Module stayed attached to the Service Module until just before reentry into the atmosphere, when the Service module was commanded to reenter the atmosphere independently and burn. The Service Module relied on the AGC in the Command Module for computation.

The lunar excursion module allowed a two man crew to land on the lunar surface, stay for a period of exploration, and return to the Apollo Command and Service Modules in lunar orbit. It had an Apollo Guidance Computer, programmed for the different and difficult tasks of landing on the lunar surface, and later taking off from the surface. Compared to the resources of the Launch complex at KSC with all its support infrastructure, the computer in the LEM did not have a lot to work with.

The LEM had two sections, one of which held the descent engine, and stayed behind on the Lunar Surface. The Ascent Stage, holding the two astronauts and the Guidance Computer, rendezvoused with the Command and Service module in lunar orbit.

At Earth launch, the Lunar Module was located between the third stage of the Saturn-V vehicle, and the command module, in the Lunar Module adapter. The command module/capsule was detached, and turned around to dock with the lunar module. When that was latched in place, the assembly was turned back to an intercept orbit to the moon, and the command module's main engine was fired. It would fire again to achieve a lunar orbit, to depart the lunar orbit after the surface exploration, and to enable reaching Earth orbit.

The iconic Apollo Guidance Computer (AGC) was developed by the MIT Instrumentation Lab, headed by Charles Stark Draper, based on the Polaris submarine-launched missile guidance computers. The Project kicked off in 1961 with a 1-page specification from NASA. The unit was designed at MIT, and was built by Raytheon

Missiles and Space Division. This was an overwhelmingly difficult task, with the state of the technology at the time. The integrated circuit had been invented only 2 years earlier. The early model computer used a core-transistor logic, and later models used a single type of NOR integrated circuit after October 1962. A prototype was built in 4 racks the size of modern refrigerators. Using a single type of integrated circuit building block simplified the procurement of parts, which were supplied by Fairchild and Signetics. An approach for in-flight repair by the astronauts for the guidance computer en-route to the moon was considered. A soldering iron was to be included in the Apollo capsule. Later, the reliability of the unit precluded this approach. The tricky part was re-packaging the four floor to ceiling racks of circuitry into a small box. A simple matter of detail. There were two identical AGC's one in the command module and one in the Lunar Excursion Module. They were programmed differently for their different tasks.

The first stage of the Saturn Rocket "stack" was the heavy lift stage, consisting of five Rocketdyne F-1 engines, one fixed in the middle, and four outside units that could swivel for steering and attitude adjustment. The first stage booster did not incorporate active guidance. The stage's job was to get the rocket and its payload from a standing start to 67 kilometers up, 93 kilometers downrange, and moving at 2,300 meters per second. That required 168 seconds of engine burn time. The total thrust developed by the engines was 7,600,000 pounds-force. Most of the first stage was fuel. The dry weight was about 130 tons, and the fueled weight was 2,300 tons. Any deviation of the vehicle during first stage burn was noted, and adjusted for during the second stage burn.

The engine's sequence of events was controlled by an onboard sequencer. This was not a computer, but just a fixed series of recorded commands that were played out in time sequence. The center engine of the stage was started 8.9 seconds before launch, with pairs of outboard engines starting at 300 millisecond intervals. This technique was used to reduce structural loading on the rocket. When the computer in the Instrument Unit confirmed thrust level correctness, the pad hold-down arms released the rocket. In the Instrument Unit, the Saturn Emergency Detection System (EDS) inhibited engine shutdown for 30 seconds after launch. It was calculated that this was safer than having a shutdown early in the sequence, which would result in a non-survivable event for the astronauts.

The sequencing of events took place on a prearranged timeline. As the vehicle lifted past the tower, it was yawed 1.25 degrees away from the tower, to provide a margin of safety in high winds. Past 400 feet, a pitch program kicked in, having been adjusted for the expected winds that month. The vehicle also rolled to the correct flight azimuth. The outboard engines were tilted to the outside, so their thrust vectors went through the vehicles center of gravity. This was to minimize the effect of one outboard engine failing. At roughly 1 minute into the flight, the vehicle broke the sound barrier. Guidance adjustment was provided by the computer in the Instrument Unit (IU). The initial trajectory was designed to gain altitude quickly as the main goal. The engines thrust grew from 7.5 million pounds-force at launch to over 9 million, in the thinner air. At the same time, the mass of the vehicle went down dramatically, as fuel and oxidizer was burned at the rate of 13 tons per second. The maximum acceleration was reached in over two minutes, at 4 G's. At this point, the center engine was shut down to limit acceleration, and the four outer engines used the remaining fuel and oxidizer. When oxidizer or fuel depletion was sensed at the pumps, the first stage was separated from the vehicle. Up high and moving fast, the first stage was separated, and the rest of the vehicle headed for Earth orbit, as the first stage fell into the Atlantic.

There were no failures of the Saturn vehicles in all of their flights, a tribute to the engineering prowess and attention to detail of the von Braun team. In the one incident on Apollo-13 (see the movie), backups and redundancy saved the day.

Saturn-V displays are located at Kennedy, Houston and Marshall. In addition, Saturn-V first stage engines, the F-1 can be seen at the Smithsonian National Air and Space Museum, Washington, DC; US Space and Rocket Center, Huntsville, A; NASA Johnson Space Center, Houston, TX; Kalamazoo Aviation History Museum (Air Zoo), Kalamazoo, MI; New Mexico Museum of Space History,

Alamogordo, NM; and the Powerhouse Museum, Sydney, Australia.

Ares

Ther Constellation was to replace the Space shuttle, and take over its duty's. Two booster rockets were in design, Ares-I and Ares-V. Ares-1 would launch crew to orbit. The larger Ares-V, was a heavy lift, un-crewed craft. Its life capacity was to be 188 metric tons to LEO. (Compare to Saturn V's 118 metric tons). In addition, the program included the Orion capsule, an Earth Departure stage, and a lunar lander. There was also consideration of a crewed mission to a near earth asteroid. The crowning achievement would be a crewed visit to Mars.

The Constellation Program was canceled in 2010, after the Senate Committee decided the program could not be executed without substantial additional funding, It was shelved by President Obama in 2010, In 2011, NASA adopted the design of a new Space Launch System.

Orion

The Orion capsule is the next generation 4-person crewed space flight unit, designed to hold four astronauts. A mock-up of the capsule is at the Kennedy Visitor's Center. It is like an expanded Apollo Capsule, an exploration vehicle designed to carry 4 Astronauts for extended missions. It is awaiting a new generation of launch vehicle, the Ares Space Launch System (SLS) to be able to carry out its mission of crewed exploration beyond low Earth orbit. The capsule is known as the

Crew Exploration Vehicle, which is larger but weighs less than its Apollo predecessor. There is a Service module with propulsion and expendables, like the Apollo design. For lunar missions, the Lunar Surface Access module would be launched separately, and rendezvous with the command and service module in orbit. The crewed portion is designed for 21 days of active time for the crew, plus an additional 6 months quiescent time. Good for the Moon, maybe an asteroid mission, not quite for Mars. An additional piece, the Deep Space Habitat is in the planning stage. It was flight tested in 2014, The first crewed flight of Orion is scheduled for 2021.

The Crew Exploration Vehicle is currently being built by a Lockheed-Martin Team. Honeywell is supplying the computers. The three flight control modules are derived from the Boeing 787 avionics project. This architecture consists of multiple computer systems linked in a network. The initial missions of Orion will be flights to the International Space Station for crew substitution and resupply of consumables. The Orion capsule is reusable, and supports a crew of four. A first, un-crewed launch was successful in 2014 by a Delta-IV Heavy vehicle out of Launch Complex 37 at Cape Canaveral. The capsule was recovered at sea. The Orion is anticipated to be the basis for a series of lunar, and later, Mars missions. The pressurized volume is 691 cubic feet, with a habitable volume of 316 cubic feet. The capsule has 50% more volume than Apollo. This is partially due to the fact that the interior controls and equipment are smaller than those of the Apollo-era. The Orion will feature an auto-dock capability. Most importantly, it includes a toilet. It uses a nitrogen-oxygen atmosphere at sea level or reduced (10 psi) pressure. It includes parachutes, and is designed for a water landing.

The Orion avionics has a networking architecture, multiple redundant units, and is radiation-hard. The Vehicle Master Computer is an adaption of Honeywell's Flight Computer for the 787 aircraft. This is based on the IBM PowerPC 750FX. This unit was introduced in 2002. For radiation hardness, it is manufactured in silicon-on-insulator (SOI) technology, and has 39 million transistors and it consumes 4 watts of power, operating at 800 MHz. In commercial versions, it was used on the Apple iBook G3. Each Flight Computer has 2 processors. The avionics components come from Rockwell-Collins. Communication between sensors, the computer, and data storage is by ethernet. A special variation, time-riggered ethernet, is used to guarantee time of arrival of critical commands.

Space Stations

This section presents an overview of Space Stations, crewed, and visited by numerous logistics flights. In more detail, this topic is covered in a companion volume from the author.

Manned Orbiting Lab

The Manned Orbiting Lab was a USAF space program circa 1963. It was to be launched into orbit, and then two astronauts could visit it via a Gemini Capsule. Mission duration of up to 40 days were envisioned. The project was a follow on to Dynasoar. Hardware was built, but the mission was never flown. There were to be seven flights, 5 of them crewed. MOL astronauts in three groups were chosen from Air Force, Navy, and Marine Corps Personnel. The project was canceled in 1969. NASA's Skylab mission was the follow-on to a crewed lab in orbit. A Gemini-B Capsule for MOL is at the U.S. Air Force Museum, Dayton, Ohio It has a circular hatch in the heat shield for entering the MOL.

It would have been launched on an upgraded Titan vehicle. The capsule would dock with the MOL unit in-orbit, and at the end of the mission, would undock, and return to Earth. It was designed to operated in Polar orbit, which would involve a launch from Vandenberg Air Force base in California. The reaction control system was different from the Gemini's OAMS.

A MOL mock-up and a refurbished Gemini capsule (not crewed) were launched from Cape Canaveral on a Titan-IIIc in 1966. The Gemini capsule reentered and was recovered. The MOL released three satellites, once it achieved orbit. Proposed first crewed flight was to be in 1970. NASA offered the military crews the opportunity to transition to the civilian space corps, and the 7 that were eligible did so. All flew on shuttle missions.

The MOL was to be built by Douglas Aircraft. In 1969, the project was canceled, since it had been demonstrated that un-crewed reconnaissance satellites could spy on our enemies much more cost-effectively.

Skylab

By the time that the Saturn rocket had put men on the moon and returned them safely multiple times, the Apollo computers were mostly obsolete. This was due to major advances in hardware and software, largely driven by the Apollo effort. With some spare Saturn's sitting around, the next project was the Skylab space station. This used a Saturn S-IVB upper stage as the structure for the station, launched by a Saturn-V with live first and second stages. The payload to orbit weighed 170,000 pounds. The station was 82 feet long, 56 feet wide, and 36 feet in the other direction. It was quite visible from Earth, when the arrays caught the glint of the Sun. Astronauts were carried to the facility in-orbit on three missions in 1973-1974 by Apollo command and service modules launched on Saturn-Ib vehicles. A second Saturn-1b and Apollo stack was kept in standby in case a rescue mission was needed. It's orbit was a near-circular 270 miles, with a 50 degree inclination. It was competing for funds and resources with the USAF's MOL project.

The facility had been damaged during launch when the integral micrometeriod shield torn away, taking one of the solar array panels with it. This caused thermal and power problems. During an un-anticipated EVA, the crew rigged a replacement heat shield, and freed the solar panel. The third crewed expedition set a record for days spent in space, at that time, 84 days.

The Industrial design firm for Skylab was headed by famed architect Raymond Lowey. He emphasized habitability and comfort for the astronaut, He included a wardroom space for meals and relaxation. He also wanted a window to view Earth and space. This has proved to be a great feature on the current ISS. Astronauts who participated in Skylab planning were dubious about the designers' focus on areas such as color schemes and decoration. They vetoed an entertainment center. Skylab food was improved over the earlier lunar mission food.

Liquid waste was not recycled as it is on the current ISS. Liquid and solid waste went into the other large tank in the facility, and was stored. It burned when Skylab re-entered.

The facility provided living and working space for the astronauts, a true shower and toilet, a solar observatory, and fixtures and services for science experiments. Some experimental data including film was brought back with the astronauts. The facility had two docking ports, and an airlock. When an Apollo capsule was docked, electrical power from the fueled cells in the service module could be used to augment power from the solar arrays.

The first Skylab mission lasted 272 days, not all occupied, followed by an uncrewed period of 394 days, when the computer kept things going. The computer was turned off for 4 years while NASA discussed reboosting Skylab to a higher orbit, or letting it reenter. There was a need to put some mods in the software, but the tools and card decks containing the code had been discarded. This resulted in some 2500 cards being re-punched from listings. At the end of 4 years, the onboard computer was booted up by ground command, and the updates worked fine. There were three total visits by 3 person crews.

There were plans to use the Shuttle to repair and reboost Skylab, but the timing did not work out. Skylab was in orbit until 1979, when it reentered the atmosphere, splashing into the Pacific ocean near Perth, Australia.

As follow-ons to Skylab, a number of proposals were presented. Von Braun envisioned a larger station, built from the second stage of the Saturn-V. The second stage would be used for fuel and liquid oxygen on the way to orbit, then vented to space, and an equipment module would be slide into the vented hydrogen tank. Questions about how many Saturn-V's would be available led to the choice of the S-IVB option.

Skylab-II

Skylab-II was a circa 2013 concept from the Marshall Space Flight Center's Advanced Concepts Office. It would be the same concept as the original Skylab, but used the upper stage hydrogen fuel tank from the Space Launch System under development. It is to be located at the Earth-Moon L2 point (Lagrange point, a null in the gravity field). Here, it would need minimal orbital adjust to remain at that spot. That particular point is on the other side of the moon, from the Earth. That puts it some 430,000 km from Earth, and 62,800 km from the lunar surface. With the moon between the station and the Earth, a relatively quiet radio environment is achieved. The goal is to support a 4 person crew for 60 days, without a resupply flight. The re purposed tank would have a diameter of 8.5 meters, larger than the ISS' 4.5 meters. The provided volume would be about 500 cubic meters. Lessons learned from the ongoing ISS mission would be applied to the Skylab-II project.

Shuttle-Mir Program

The Shuttle-Mir program was a demonstration of cooperation in Space, and led to the combination of two space station programs of the United States and Russia into one facility, the International Space Station. The U.S. wanted to take advantage of the Russian on-orbit experience, and Russia needed hard cash. This was a successful joint endeavor, continuing to this writing, regardless of the motivations.

Shuttle mission STS-63 flew to the MIR station in 1995. There was no attempt ot dock; only a fly-around. There would be a total of nine Shuttle flight to Mir, delivering new modules for station expansion, crew exchange, and logistics supplies. American astronauts logged close to a thousand days onboard MIR. The station was a follow-on to the earlier Salyut Stations. It was larger and more comfortable, and took advantage of lessons-learned in orbit. With the docked Shuttle, the facility massed 250 metric tons, the world's largest facility in space.

The MIR Station was de-orbited successfully in March of 2001. It had flow three times its projected life.

China

The Shenzhou spacecraft resembles the Soviet Soyuz, but is 10% larger. First launch was in 1999, with two additional uncrewed flights first crewed flight in 2003.

There are three sections, the orbital module, the re-entry module, and the service module. There was no docking module. The orbital module stays in orbit, can dock with a later spacecraft.

The Tiangong Space Station finished construction in 2022. It is scheduled to become pwermanently occupied in 2023, with a proposed operational phase until 2038.

International Space Station – ISS

We will just touch upon the permanent International Space Station, but it will be covered in detail in a follow-on book specifically about space stations from the author.

In 1993, United State's Space Station Freedom Project,which became the International Space Station, kicked off. On-orbit construction began in 1998, and was completed with a last Shuttle mission in 2011. It is the largest artificial satellite in Earth orbit, and can be seen from the ground with the naked eye. The ISS is a synthesis of several space station projects from America, the Soviets/Russians, the Europeans, Canada, and the Japanese.

The International Space Station is continuously crewed, and orbits the Earth at an altitude of some 250 miles. It is quick, traveling at 17,300 miles per hour. It is also expensive, representing an investment of some $100+ billion dollars by the world community, mostly by the United States and Russia. It is thus the most expensive object ever constructed by mankind. It has been visited by astronauts and cosmonauts from some 15 nations, and by paying tourists.

The onboard computing infrastructure was based on rad-hard Intel 80386 32-bit computers for housekeeping. There is an wired Ethernet backbone in the Station, as well as WiFi. Applications supported include IP phone with webcam for crew conversations with family's.

The laptops replaced the original design of the Multi-Purpose Application consoles, which were MIL-Spec equipment, purpose-built and programmed from scratch using the ADA language and X-windows. The laptops (PGSC – Portable General Support Computers) were more flexible. The general concept is that COTS machines can display any data, but must follow a arm/check/fire protocol to send commands. Originally, the support computers were Grid laptops. These were 8086-based machines, running GridOS.

The Station used some 68 COTS IBM/Lenovo ThinkPad A31 laptops, 32 ThinkPad T61 laptops, and a T61 as a server, with routers. The laptops handle non-critical and experimental support as well as music and television for crew entertainment. The computers run Windows-XP, with some 3 million lines of flight code. The architecture was designed as a station-wide distributed system.

The IBM/Lenovo laptops were modified to handle the environment of the station. With no convention cooling, due to lack of gravity, fans had to be added to keep components on the motherboard cool. In addition, the circuit boards and connectors were conformally coated to contain any debris. The power supplies were modified to accept Station 28 vdc power. In addition, the laptop had to operate in the 10 psia atmosphere of the Space Shuttle, even though the Station maintains a sea level pressure of 14.7 psia. The laptops were also modified to have Velcro for attaching to convenient surfaces.

The NASA ISS ap for Android is great. It shows the point on the Earth the Station is above at the time, and has a high resolution camera that returns images (during Sun-lit portions of the orbit).

The crews are rotated in and out of the station for varying periods, not to exceed a year. There is always one capsule docked at the station for emergency evacuation.

Logistics flights (food, oxygen, science projects, clean underwear) up, and trash down use a variety of options, the Soyuz cargo version, and commercial uncrewed capsules under contract to NASA. Some capsules burn in the atmosphere upon reentry, and some can be recovered and reused.

Crew Exchange is done currently only by the Soyuz TM. Resupply is by Russia's *Progress* (15000 lbs), Orbital-

ATK's *Cygnus (*4400-7700 lbs, depending on launch vehicle*)*, and SpaceX's *Dragon* (6400 lbs), the European Automated Transfer Vehicle (19,500 lbs), and the Japanese HTV Transfer vehicle (9900 lbs).

NASA's Commercial Crew Development Program (CCDev)

The CCDev Program is part of NASA's Commercial Crew and Cargo Program Office (C3PO), that is designed to stimulated the development of privately operated crew and cargo vehicles to low Earth orbit. This will help fill the gap left by the decommissioning of the Shuttle fleet in 2011. The program started in 2010. In 2014, contracts were awarded to SpaceX and Boeing. The high level requirements are 1) to deliver and return 4 crew members and their equipment to the ISS; 2) to provide crew return in case of an in-space emergency; 3) to serve as a 24-hour safe haven in space, in case of an emergency; and 4) be capable of spending 210 days in space, docked to the ISS. At the current time, the first launch of a key component, the Space Launch system (SLS) is scheduled for no earlier than 2021.

NASA's cost for a seat to orbit in the Soyuz vehicle is a little more than $70 million. Under the program, an un-crewed test flight of the SpaceX Dragon 2.0 is scheduled for February 2018, followed by a crewed flight shortly after. The Boeing CST-100 Starliner is scheduled for similar flights in June 2018.

Soviet Union/Russia

As with the United States, the Soviet Union used German World War-II rocket technology as a basis. The U.S. had von Braun, and the Soviets had Sergei Pavlovich Korolev, a brilliant designer and engineer. He set the direction for the country, which went on to put the first artificial satellite in orbit (1957) and the first man in orbit (1961). The systems were tested in a series of launches of dogs going into Earth orbit as precursor missions. Officially, the Soviet Union (Union of Soviet Socialist Republics) became the Russian Federation in 1991.

Vostok

Vostok ("East" in Russian) was a Soviet era capsule holding one cosmonaut. The first man in space, Yuri Gagarin, made his flight in a Vostok capsule in 1961. Instead of a water landing as the United States chose, the capsule made a landing within the borders of the the-Soviet Union. The Cosmonaut parachuted from his capsule, and landed separately. The payload to orbit weighed 4,725 kg. The descent and reentry portion of the capsule was spherical. It separated from the instrumentation section after the descent engine fired. At seven kilometers from the surface, the cosmonaut ejected from the capsule, and parachuted to the ground. The flight was repeated 4 months later by Cosmonaut Titov. Thirteen crewed flight were scheduled for the Vostok. Only six were conducted, as the Voskhod Program was phased in. Valentina Tereshkova, the first woman in space, rode Vostok 6 in 1963. The longest flight was made by Valery Bykovsky, nearly 5 days in space. Vostok 3 and 4 were in orbit simultaneously, in 1962

The basic Vostok design was resilient, being used for nearly 40 years. It consisted of a spherical descent module, 2.3 meters in diameter, and a someway conical instrument module. An ejector seat served as a way for the cosmonaut to exit rapidly in case of a launch vehicle problem. It was a viable option for the first 40 seconds of flight. There were no viable options during the first 20 seconds after ignition. Chief Designer Korolev made his spacecraft more automated, so cosmonauts did not necessarily need to be qualified pilots. The 3KA model Vostok was used in the initial flights. After uncrewed flights, 6 were launched with a crew member.

The equipment module massed 2,270 kilograms, and was 2.25 meters long by 2.4 meters in diameter. It had the main engine for reentry, and reaction control thrusters for attitude adjustment. It could carry expendables for a 10-day stay in orbit.

Voskhod

Voskhod ("sunrise" in Russian) was a Soviet-era program, and capsule, a follow-on to Vostok, capable of hosting three. The Vostok design was augmented by a solid fuel retro rocket, because the Cosmonauts rode the capsule to the ground. The ejection seat was removed. There were two flights, in 1964 – 1965. It was designed by the legendary Sergei Korolev. The Voskhod program transitioned to the Soyuz in 1967. Voskhod used a Vostok spacecraft with a solid fuel retro rocket for the descent module. There was not enough interior space for three cosmonauts to wear spacesuits. In the two-crew member configuration, they could. This was used for missions requiring EVA. There were two variations, one for 2 crew members in suits, and one for three crew members without suits. There was no launch escape system.

The Voskhod-2 mission included an airlock, and Cosmonaut Leonov took advantage of this to step outside for a while. The capsule and equipment was air-cooled, so it needed the air lock. After use, the air lock was jettisoned.

The Voskhod-3 carried three, without space suits. There were two flights. It could handle the 3-person crew for 14 days on orbit. The reentry module was a 2.3 meter sphere. The equipment module was 2.3 meters in diameter, and 2.4 meters long. It massed 2300 kg. The 3KD version went back to a crew of three. It flew twice in 1965, once empty and once with a crew.

The spherical descent module had a diameter of 2.3 meters. The conical equipment module massed 2.2 (metric) tons, and was 2,25 meters long and 2.4 wide. The equipment module housed the rocket engines for reentry.

Soyuz

The Soyuz (Russian: Union) program was started in the early 1960's with a goal of launching a Cosmonaut into Earth orbit. It involved the development of the rocket as well as the space capsule. Derivative Soyuz launch vehicles are still in use, in International Space Station resupply missions, as well as crew exchange.

The original Soyuz capsule had three sections: the orbital module, which was a sphere, the reentry module, and the service module, with expendables and solar panels. The orbital module and the service module are single use. The consumables allow for 30 person-days in orbit, before reentry. The current uncrewed *Progress* cargo ship for the ISS is a derivative of the Soyuz. The first flight was uncrewed, and was accomplished in November, 1966. Currently, a Soyuz remains docked to the ISS at all times, in case the inhabitants have to evacuate. The Soyuz life in orbit is 6 months.

The orbital module serves as a habitation module on-orbit. It includes a toilet. The internal volume is 6 cubit meters. The orbital module can be customized on a per-mission basis. It must be jettisoned from the capsule before reentry.

The descent module, the inhabited capsule, is meant for re-entry, and has a large parachute, and a heat shield. There are also braking rockets. It is spherical in shape, as this gives the best volume, four cubit meters. The crew usable volume is 2.5 cubic meters. The crew capsule contains the airlock allowing entry to the service module,

The service module is cylindrical and has both pressurized and non-pressurized sections. The service module hosts the reentry and maneuvering main engine, and a solar array. In several cases, an emergency situation occurred when there was an incomplete separation between the service module and the descent module.

The next two Soyuz flights had a single cosmonaut onboard. Soyuz missions 5, 7, and 8 carried two. The rest of the missions through Flight 10 carried three cosmonauts. Missions 11 (1973) through 66 (1991) generally carried three, with some exceptions with 2 crew onboard. After that, the flights were conducted by the Russian Federal Space Agency. This entity has overseen flights 66 (1992) through 134 (July 2017).

All Soyuz missions are launched from the Baikonur Cosmodrome in Kazakhstan. It has a remarkable operational history, in that the only other flight to suffer a fatal accident was Soyuz 11, where the crew died when the cabin pressure was lost just before reentry. Soyuz vehicles have serviced three space stations.

The Soyuz assemblies have evolved and been updated over their service life. The 1963 Soyuz-A massed 5,880 kg and was 7.4 meters long. The new TMA version masses 7,220 kg, and is slightly larger at 7.48 meters. The orbital module has gone from 1,000 kg with a volume of 2.2 cubic meters to 1,270 kg, and 5 cubic meters. The reentry module went from 2,400 kg to 2,900 kg,

The Soyuz has gone through four generations, with the latest being the MS model. It has a much improved approach and docking system, with a new TsVM-101 onboard computer, which is one eighth the weight of the original Argon-16.

Buran, the Russian Shuttle

Buran ("snowstorm" in Russian) was a Russian spaceplane, and the name of the project. It completed one 3 hour+ non-crewed flight (2 orbits) in 1988. It was commanded by ground control to de-orbit and landed at the runway at the Baikonur Cosmodrone in a strong cross-wind. It was tragically destroyed in its hanger, when the roof collapsed due to snow load in 2002. The collapse also killed eight workers. It resembled the U. S. Space shuttle, but included the auto-land mode.

Buran had a sophisticated system of dual 4-computer units, with voting logic. Unlike the Shuttle, the Buran could be flown and recovered in an unmanned mode.

Like the Shuttle, the Buran was transported to the launch site on the back of a large aircraft, the Antonov AN-225 *Mriya (*Russian: Inspiration*")*. That plane is the longest and heaviest ever built, and uses six turbofan engines. The maximum take-off weight is 640 metric tons. A second aircraft was partially constructed, and later finished for delivery to China. The aircraft has 32 wheels on its landing gear, and a crew of six.

Joint Missions

This section discusses space mission that involved multiple crewed spacecraft from different nations joining up in orbit.

Apollo-Soyuz – the match-up

The Apollo-Soyuz Test Project was a joint space project between the Soviet Union and the United States to fly a similtaneous missions that would meet up and join on-orbit. It was conducted in July, 1975. It required a special docking adapter that allowed the two different space capsules to join up. You can see a coupled Apollo-Soyuz at the Smithsonian Air & Space Museum in D. C. The Apollo is the unit used for pre-flight testing, and the docking module between the craft is the back-up model, made by North American Rockwell. A similar display is at the Cosmosphere, in Hutchinson, Kansas. The actual Apollo capsule from the flight is on display at the California Science Center in Los Angeles. If you want to see Soyuz-19 that made the flight, you will have to go to the RKK Energiya museum in Korolyov, Moscow Oblast, Russia.

The Apollo was the last of its kind to fly, as it was replaced by the STS- Space Shuttle for further crewed missions. Despite the technical and ideological differences, the mission was a great success, with real science being conducted.

At the time of the joint flight, tensions between the then-Soviet Union and the United States ran high. This successful project was proof of "detente" and cooperation.

Millions around the world watched the event on live tv, because the author caught the red-eye back from JPL to

GSFC, and fixed the onboard computer of the communications relay satellite (ATS-F).

Chinese Missions

This section discusses the Chinese crewed spacecraft program.

Shenzhou

The Chinese Shenzhou ("Divine Craft") generally follows the design of the Soviet/Russian Soyuz, which was tech- transferred by the Russians. The deal involved training two Chinese astronauts at the Yuri Gargarin Cosmonaut Training Center The program successfully put the first Chinese citizen, Yang Liwei, into orbit in in Shenzhou-5 in October of 2003, after four un-crewed test flights. It has a design life on orbit of 20 days, limited by expendables. At launch, it masses 7840 kg. It has a volume of 14 cubic meters.

The Shenzhou spacecraft is modeled on the Soyuz, but is larger. Like the Soyuz, the configuration includes an orbital module, a service module with solar cells, and a reentry capsule. Other than the reentry module, everything else stays in orbit. The reentry module is single use. It uses a single parachute, and braking rockets, like Soyuz.

Through October of 2016, there have been eleven Shenzhou flights, 6 of these crewed. The first Chinese woman in space flew on the 9th mission. The missions progressively demonstrated rendezvous and docking, and spacewalking. The last two missions docked with the Tiangong-1 and -2 space stations.

ESA

CSTS (Crew Space Transportation System) or ACTS (Advanced Crew Transportation System) is a human spaceflight system proposal. It was originally a joint project between the European Space Agency (ESA) and the Russian Space Agency (FKA), but became solely an ESA project. NASA is pursuing their next crewed spacecraft, Orion, without foreign participation. ESA envisions the CSTS being used in lunar missions. This would involve multiple launches, and rendezvous and docking in orbit. The Ariane rocket could be used, if updated to human-rated standards.

ISRO

The Indian Space Research Organization is working toward a crewed spacecraft that will support 3 astronauts for seven days in Earth orbit. A simple version, proof-of-concept, is planned with rendezvous and docking capability to be added later. The Indian Human Spaceflight Program kicked off in 2007. The first uncrewed flight is scheduled for 2023; it is called Gaganyaan 1. Thei are looking forward to a space station, and crewed lunar landings.

The Orbital Vehicle will be a capsule of 3.7 metric tons, holding three, and capable of remaining in orbit seven days. The launch vehicle, the GLSV Mk III, for this craft is currently under development. It will be somewhat larger than the U.S. Gemini, but not as large as an Apollo or Soyuz.

Previously, ISRO had launched an 1,200 lb uncrewed capsule to orbit in 2007. It reentered in 12 days, and splashed down in the Bay of Bengal. It used a silica thermal tile heat protection system, and a parachute. Not just a proof of concept, several experiments were conducted onboard during orbit. A second launch demonstrated re-entry and recovery procedures.

Private efforts

This section discusses non-Governmental human space projects. In 1966, for the Ansari X-Prize of $10,000,000. to "the first non-government organization to launch a reusable crewed spacecraft into space, twice in two weeks" was announced. Twenty-six teams from around the world competed. The prize was collect in October of 2004, 5 months later, by a craft designed by Burt Rutan, and financed by Paul Allen of Microsoft. This competition seems to have had the result of getting a lot of rich people interested in space. Private companies are fulfilling the logistics role for the ISS since the demise of the Shuttle program.

Blue Origin

"...to build space hotels, amusement parks and colonies for 2 million or 3 million people who would be in orbit. 'The whole idea is to preserve the earth' he told the newspaper .... The goal was to be able to evacuate humans. The planet would become a park." Jeff Bezos, High School Class valedictorian, 1982, Miami Herald.

Blue Origin is a privately funded spaceflight services company founded by Jeff Bezos, of Amazon and the Washington Post Company. It was founded in the year 2000 in Kent, Washington. The company "is developing technologies to enable private human access to space with the goal to dramatically lower costs and increase reliability." They are focusing on vertical take-off, vertical landing reusable vehicles. The name refers to Earth. Their spacecraft is named *New Shepard,* and has been flown uncrewed as of this writing. It went almost to space, reaching just shy of 100 km. The second flight did go beyond 100 km, and both the capsule and booster were recovered. They have re-flown the hardware from the first mission several times. Crewed flight will happen sometime in 2018. The company uses its own launch vehicles. The company is located in Washington State, and uses a launch facility in west Texas for sub-orbital launches. They have leased facilities at the Kennedy Space Center for orbital launches. Some technology from the earlier McDonald Douglas-NASA DC-X project influenced the Blue Origin. They focus on reusable vehicles.

The orbital rocket is named "N*ew Glenn*" after Astronaut John Glenn, The first stage jas seven of their BE-4 engines, and will be recoverable and reusable. The BE-4 engine production facility is in Huntsville, AL. The rocket portion was recovered and reused a total of three times. Their engine is the BE-4, a liquid oxygen/liquid methane design. It is unfortunately, at this writing, some 4 years late. It was to be used by United Launch Alliance.

The Crew capsule of the New Shepard will hold three or more astronauts, and has been launched, uncrewed, to space, by the definition of 100 km. They use a private spaceport in Texas.

The company has a NASA Space Act Agreement in place under the Commercial Crew Development Program.

Virgin Galactic

Virgin Galactic is providing suborbital flights, and plans to support space tourism as well as science missions. The company is owned by Sir Richard Branson. It was the first privately built and operated vehicle to achieved crewed spaceflight. This was on the 100th anniversary of the Wright Brother's flight, From airplanes to space plans to orbit in 100 years. Let's keep that up.

Scaled Composites, owned by Virgin Galactic, developed a air-launched vehicle, for access to space for tourists. The carrier vehicle, White Knight Two, carries the smaller SpaceShip-2 to a high altitude,. At this point the two vehicles separate, and the White Knight flys back to a runway landing as the smaller craft continues to orbit with its rocket engine. The two vehicles were designed and built by aviation and space pioneer Burt Rutan. The plan is to have five SpaceShip-2's, the orbital vehicle, and two WhiteKnight-2's, the crewed carrier vehicle.

Two SpaceShip-2's are under construction, and there is one White Knight-2, the carrier ship.

The White Knight, sometimes called a Flying Space Aircraft Carrier, has a crew of two. It's cargo capacity is 37,000 lbs to 50,000 feet. If it carries an "upper stage", the Launcher-one, it can put 200 kilograms to low Earth Orbit. White Knight has four Pratt & Whitney turbofan engines. The wingspan is 41 feet, and there are dual fuselages. The second fuselage can hold additional crew members, or tourists.

Unfortunately, the first SpaceShipTwo broke up during a test flight, killing one pilot, and injuring the other, although he parachuted to the ground. Before the accident, the vehicle did indeed reach "space." The second unit is undergoing flight testing at this writing.

It glides back to a landing at Spaceport America in the Mojave Desert.

SpaceX

Space-X has an impressive track record, including the first privately funded liquid rocket engine to orbit, the first to launch and recover a rocket, and the first private company to send a vehicle to the ISS, and the first reuse of an orbital rocket. Their Dragon capsule is designed in two versions, to take personnel or supplies to the ISS. The cargo version has been in operation since 2010. After the newer version is declared operational, the old version will still be used for cargo flights.

Their newer Dragon V2 capsule has a capacity of 7 crew. The Dragon capsules have an integral launch escape system, consisting of rocket engines. The vehicles are capable of autonomous rendezvous and docking. The onboard controller is a tablet computer. They can parachute to a water landing, or do a dry landing, assisted by the external rocket motors. The capsule is partially reusable, with a projected life of ten flights before refurbishment.

The target launch prise was $160 million. In a crewed version, this works out to $20 million per seat. Soyuz charges $76 mission per seat. All personnel wear space suits while in the capsule.

Space-X's privately funded Falcon rockets are reusable. First launch was in 2008, setting a milestone for privately funded liquid rockets. The first three launches were failures, but the company analyzed the situation, and made adjustments. The rockets are recovered on drone ships and re-used. In 2018, 100 launches were completed. The rocket usually launches from Kennedy's Launch Complex 39A.

The Merlin engines use liquid oxygen and kerosene (RP-1). The Drago engines use the hypergolic hydrazine and nitrogen tetroxide. The Raptor series use liquid methane and liquid oxygen.

The Dragon-9 launch vehicle is recoverable and reusable. The European Ariane rocket was falling behind in 2014, and costs were adjusted.

The Falcon-1 vehicle was retired in 2009. The Falcon-9 and Falcon-9 heavy life are in service. Falcon-9 has 27 of the company's Merlin engnes. Falcon Heavy launched Mr. Musk's personal Tesla Roadster into a helio-cemtric orbit in 2018.

As of February 2022, the Falcon series has launched 184 times. There were 140 successes one partial and one failure. They launch from Kennedy Space center in Florida, Vanderburg Air Force Base in California, and their facility in Boca Chica, Texas.

The Space-X Super Heavy has completed five successful high altitude flights, and were recovered. The Starship spacecraft is a crewed vehicle. The entire stack is referred to as Starship. It is designed to go to the Moon and Mars. Next year, Starship is scheduled to conduct a private lunar mission.

Sierra Nevada's Dreamchaser

This craft is a 7-crew Earth orbital space plane, developed with $20 million of NASA seed money. The Dream Chaser was based on the circa-1990 NASA HL-20 Personal Launch System, a mini Space Shuttle designed as a lifting body. NASA-Langley, the lead center for the project, had not gotten to the implementation stage. The unit would have limited cargo capacity, but could get crews to and from orbit, particularly the ISS.

Sierra Nevada did not make the final cut in the competition, losing to SpaceX and Boeing, supposedly due to lack of technology maturity. Sierra Nevada decided to pursue the Dream Chaser project independently.

Sierra Nevada did win a Commercial Resupply Services contract from NASA, for a minimum of six launches. These missions will use the Atlas V launch vehicle.

Sierra Nevada also supplied the rocket motors for Virgin Galactic Spaceship Two, and for Scaled Composites.

Boeing CST-100 Starliner

The Boeing Starliner is a next-gen crewed space capsule for low Earth orbit. It could be used to exchange crews on the International Space Station. It is designed to be used up to 10 times. It comes equipped with wireless internet. NASA is seeking 6 crew rotation missions from the Starliner. The Starliner is compatible with the Atlas V, Delta IV, and Falcon 9 launch vehicles.

The Starliner can hold up to 7 crew members, or a mixture of crew and logistics supplies. The crew displays take advantage of new Aerospace practices in the glass cockpit, where tablet-based technology is used to replace traditional gauges and displays. The capsule can be flown un-crewed, and can stay in space for up to 7 months. It is somewhat smaller than the Orion capsule.

Excalibur Almaz

Excalibur Almaz is an ambitious private space flight company planning not just Earth orbit, but lunar and deep space services. It is cooperating with XCOR Aerospace. The company is located on the Isle of Man. It has a Space Act Agreement with NASA in the commercial Crew Development area. They plan to use the hull and general configure of surplus capsules from the Russian Almaz military space station program, updated with more modern electronics. The plan is to launch this station into space, and have it visited by researchers and space tourists. It remains to be seen whether this plan will continue.

Stratolaunch Systems

This space start-up combined Paul Allen, of Microsoft, with Burt Rutan of Scaled Composites. The project involves a carrier aircraft and a multistage air-launched vehicle. Their facility's are located in Mojave, California. They are focused on hypersonic flight, not so much as achieving orbital.

Space Tourists

There have been, to this point, 7 civilian space tourists, who have paid their way into orbit. If you're getting paid to do it, you're an astronaut. If you pay to do it, you're a tourist. Space Tourism will be examined in a companion volume.

- Dennis Tito was the first to pay his own way to visit the ISS in 2001 on a Russian Soyuz craft.

- Mark Shuttleworth was second.
- Gregory Olsen went to the ISS.
- Anousheh Ansari was fourth, flying in a Soyuz capsule.
- Charles Simonyi flew in a Soyuz. He had some much fun, he went a second time.
- Guy Laliberté, a Canadian, flew in a Soyuz
- Yi-So-yeon, a Korean, went to the ISS
- Christa McAuliffe, Teacher in space, was killed in Challenger disaster.
- Helen Sharman, first Briton in space, went via Soyuz to Mir.

Spaceflight participants – these flew to orbit, but are not professional astronauts.

- U. S. Congressman Bill Nelson flew on the Shuttle Columbia.
- Jake Garn, of the Senate Appropriations Committee, flew as a payload specialist in shuttle Discovery.
- Japanese TV reporter Toyohiro Akiyama flew to MIR 1990. It was a business trip. TBS
- Dr. Sheikh Muszaphar Shukor Al Masrie bin Sheikh Mustapha of Malaysia, andorthopedic surgeon, went to the ISS, and, no, it was not a house call.
- Richard Garriott, son of Astronaut Owen Garriott paid his way to the ISS, becoming the first 2nd generation space-farer.

The fallen Astronauts memorials

Spaceflight is not yet routine. It is hazardous. There is an Astronaut Hall of Fame at the Kennedy Space Center, as well as a Fallen Astronaut Memorial. There is actually a memorial to fallen Astronauts and Cosmonauts on the Moon, placed there by the Apollo-15 crew at Hadley Rille.

Afterword

I hope to again re-issue this book in a few years, with updated information and new projects to talk about. Can I wrap up the final edit in orbit? Let's see.

Bibliography

The Astronauts, *We Seven*, 1962, Simon and Schuster, 2010, Renewal Edition, ISBN-1439181039.

Baker, David *NASA Mercury - 1956 to 1963 (all models): An insight into the design and engineering of Project Mercury - America's first manned space programme,* 2017, Haynes Publishing, ISBN-1785210645.

Baker, David *Soyuz Owners' Workshop Manual: 1967 onwards (all models) - An insight into Russia's flagship spacecraft, from Moon missions to the International Space Station,* Haynes, 2014, ISBN-0857334050.

Baker, David *International Space Station: An insight into the history, development, collaboration, production and role of the permanently manned earth-orbiting complex*, Haynes, 2016, ISBN-0857338390.

Baker, David *NASA Space Shuttle Manual: An Insight into the Design, Construction and Operation of the NASA Space Shuttle*, Zenith Press, 2011, ISBN-0760340765.

Baker, David *North American X-15 Owner's Workshop Manual: All types and models 1959-1968, Haynes, 2016,* ISBN-085733767X.

Baker, David *The History of Manned Space Flight,*

1982, Crown Publishers, ISBN-051754377X.

Behrens, Carl E. *Space Launch Vehicles: Government Activities, Commercial Competition, and Satellite Exports,* 2006, ASIN-B005Y181WO.

Belfiore, Michael *Rocketeers, How a Visionary Band of Business Leaders, Engineers, and Pilots is Boldly Privatizing Space*, Harper Collins, 2008, ISBN-0061149039.

Bentley, Matthew A. *Spaceplanes: From Airport to Spaceport,* 2009, Springer, ASIN-B008BB7HQA.

Bergaust, Erik, *Werner von Braun*, National Space Institute, 1976, ISBN-0917680014.

Berger, Eric *Liftoff: Elon Musk and the Desperate Early Days That Launched SpaceX* , 2021, ASIN-B088FQK2K2.

Bilstein, Robert E. *Stages to Saturn, A Technological History of the Apollo/Saturn Launch Vehicles*, 2003 edition, University Press of Florida, ISBN 0-8130-2691-1.

Brodsky, Robert *Catch a Rocket Plane,* 2012, Foxbro Press, ASIN-B009Z51NN4.

Brooks, Courtney G.; Grimwood, James M.; Swernson, Jr. Loyd S. Dickson, Paul *Chariots for Apollo: The NASA History of Manned Lunar Spacecraft to 1969,* 2009,

Dover, ISBN-0486467562.

Buckley, James *Home Address: ISS: International Space Station,* 2015, Penguin, ASIN-00YBBKM16.

Burrough, Bryan *Dragonfly, An Epic Adventure of Survival in Outer Space,* 1988, Harper Collins, ISBN-0-88730-783-3.

Burt, Dennis *Elon Musk will Take Us to Mars: How and Why the Billionaire Entrepreneur and his SpaceX Start-Up are Making Interplanetary Travel a Reality,* 2013, ASIN-B00G0T6D6U.

Caiden, Martin *Man into Space,* Pyramid Books, May 1961, ASIN- B0007E63KA.

Caiden, Martin *X-15: Man's first flight into space,* 1959, Scholastic Book Services, ASIN-B0007FSAW8.

Cernan, Eugene *The Last Man on the Moon: Astronaut Eugene Cernan and America's Race in Space,* 2000, St. Martin's Griffin, ISBN-0312263511.

Collins, Michael *Carrying the Fire,* Ballantine Books, 1975, ISBN-0374531943.

Compton, William David *Living and Working in Space: The NASA History of Skylab,* Dover, 2011, ISBN-0486482189.

Crocker, Alan R. *Making Human Spaceflight Practical*

and Affordable: Spacecraft Designs and their Degree of Operability, NASA Johnson Space Flight Center, 2015, alc Books, ASIN-B0157AEAVM.

Dethloff, Henry C.; Dickson, Paul *Suddenly, Tomorrow Came: The NASA History of the Johnson Space Center*, 2012, Dover Books on Astronomy, ISBN-0486477568.

DoD *High Frontier: The U. S. Air Force and the Military Space Program - MOL, Dyna-Soar,* 2017, ISBN-1520775709.

Dubbs, Chris, Walker, Charles D. *Realizing Tomorrow: The Path to Private Spaceflight,* U. Nebraska Press, 2011, ASIN-B0056G5WUG.

Dunn, Joeming, Dunn, Ben *Laika: The 1st Dog in Space* (Famous Firsts: Animals Making History, 2011, ISBN-1616416416.

Edberg, Don; Costa, Wilie Design of Rockets and Space Launch Vehicles, 2020, ISBN-978-1624105937.

Ellsworth-Smith, Lincoln *A Day in the Life Aboard the International Space Station*, 2015, ASIN B00V3RNAXS.

Ezell, Edward Clinton and Ezell, Linda Neuman *The Partnership, a History of the Apollo-Soyuz Test Projec*t, (NASA SP-4209), The NASA History Series, 1978,ASIN-B0000E96NI.

Faherty, William Barnaby and NASA *Apollo and*

America's Moon Landing Program - Moonport: A History of Apollo Launch Facilities and Operations - Saturn 1, Saturn 1B, and Saturn V Rocket Launch Pads, Launch Complex 39 (NASA SP-4204), 2011, ASIN-B00688W0OW.

Farbman, Melinda, Gaillard, Frye *Spacechimp: NASA's Ape in Space,* Enslow Publishers, 2000, ISBN-0766014789.

Froehlich, Walter *Apollo Soyuz*, 1976, NASA, EP-109, ASIN-B0006CWXQQ.

Gatland, Kenneth *Manned Spacecraft,* 2nd ed., 1976, MacMillan Publishing Co., Inc. ISBN 0-02-542820-9.

Gibson, Karen Bush *Women in Space: 23 Stories of First Flights, Scientific Missions, and Gravity-Breaking Adventures,* Chicago Review Press (February 1, 2014), 2014, ASIN-B00HXZN0KW.

Glenn, John, *John Glenn, A Memoir,* 1999, Bantam, ISBN-0-553-11074-8.

Goldstein, Margaret J. *Private Space Travel: A Space Discovery Guide* (Space Discovery Guides) 2017, ASIN-B01MQTWSPS.

Goodwin, Robert *Dyna-Soar: Hypersonic Strategic Weapons System,* Apogee Books Space Series 35, ISBN-1896522955.

Goodwin, Robert *Rocket And Space Corporation Energia,* ISBN-1896522815.

Goodwin, Robert, *Project Apollo, the Test Program,* 2006, Apogee Books, ISBN- 1-894959-36-1.

Greene, John A. *Kennedy Space Center: Celebrating 50 years 1962-2012,* 2012, ASIN-B00904HVLS.

Greene, John A. *To Fly What Others Only Imagine: NASA Dryden Flight Research Center Historic Aircraft,* 2012, ASIN-B008L41376.

Godwin, John A. *X-15: The NASA Mission Reports: Apogee Books Space Series 13,* 2001, Collector's Guide Publishing Inc., ISBN-1896522653.

Green, William, *Rocket Fighter (German Komet fighter),*1971, 1st ed, Ballantine Books, ISBN-0345258932.

Griehl, Manfred *German Rocket Planes,* 2000, Greenhill Books, ISBN-1853674044.

Hacker, Barton C. and Grimwood, James M., Greene, John A. *On The Shoulders of Titans: A History of Project Gemini* (Annotated & Illustrated) (NASA History Series), 2012, ISBN-B008SGT1P8.

Hallion, Richard P.; Gorn, Michael H. *On the Frontier:*

Experimental Flight at NASA Dryden, 2003, Smithsonian, ISBN-1588341348.

Hendrickx, Bart; *Energiya-Buran: The Soviet Space Shuttle*, 2007, Springer-Praxis, ISBN 978-0-387-69848-9.

Hilliard, Richard *Ham The Astrochimp,* 2007, Boyds Mills Press, ISBN-1590784596.

Isakowitz, S J. Hopkins, J. *International Reference Guide to Space Launch System,* 2014, ISBN-978-1563475917.

Krantz, Gene, *Failure is Not an Option*, 2009, Simon & Schuster, ISBN-1439148813.

Lardas, Mark, Palmer, Ian *Space Shuttle Launch System 1972–2004,* New Vanguard, 2012, ASIN-B01DPPWI74.

Ladwig, Alan "The SpaceFlight Participant Program - Taking the teacher and classroom into space," 1985, NASA Technical Reports Server. document ID no. 19860031168.

Lardier, Christian *The Soyuz Launch Vehicle: The Two Lives of an Engineering Triumph*, 2013, ISBN-978-1461454588

Ley, Willy and Aldrin, Buzz, *Rockets, Missiles, and Men in Space: A Definitive Account of the History of Space,*

1966, Viking Press, ASIN-B004H3G0WI.

Ley, Willy and Von Braun, Werner, *The Exploration of Mars*, 1956, Sidgwick & Jackson; First Edition, ASIN-B0000CJKQN.

Ley, Willy, Rockets, *The Future of Travel Beyond the Stratosphere*, 1945, Viking Press, ASIN- 0007E7IC2.

Linehan, Dan, Clarke, Arthur C. *SpaceShipOne: An Illustrated History*, 2011, Zenith Press, ISBN-0760339880.

Low, George M., *Apollo Spacecraft*, NASA Manned Spacecraft Center, Houston, TX. Available in .pdf format: http://klabs.org/history/papers/low_69.pdf

Miller, Ron *Spaceships: An Illustrated History of the Real and the Imagined*, 2016, Smithsonian Books, ISBN-1588345777.

Mullane, Mike *Riding Rockets, the Outrageous Tales of a Space Shuttle Astronaut*, 2007, Scribner, ISBN-0743276833.

Murray, Charles; Cox, Catherine Bly *Apollo,* South Mountain Books, 2004, ISBN-0976000806.

NASA, *America's Spaceport: John F. Kennedy Space Center,* 2014, ASIN-B00L82609O.

NASA, *Inside the International Space Station (ISS): NASA International Space Station Familiarization Astronaut Training Manual - Comprehensive Review of ISS Systems*, 2011 ASIN-B006O403MG.

NASA, *Reference Guide to the International Space Station*, 2014, ASIN-B00M3K6LP8.

NASA, *Project Mercury Familiarization Manual, Manned Satellite Capsule*, 2011, ISBN-1935700685.

NASA, *Skylab Mission Report: Saturn Workshop, Marshall Space Flight Center - Technical and Engineering Details of Station Hardware, Subsystems, Experiments, Missions, Crew Systems*, 2012, ASIN-B00870HEBS.

NASA, *Orion: America's Next Generation Spacecraft - A Look at the Design, Development, and Testing of NASA's Multi-Purpose Crew Vehicle (MPCV) for Deep-Space Manned Exploration Flights*, 2011, ASIN-B00696ITBG.

NASA *The Apollo Spacecraft, a Chronology*, SP-4009, in 4 volumes,
avail: https://www.hq.nasa.gov/office/pao/History/SP-4009/cover.htm

NASA, *NASA's Constellation Program: Lessons Learned (Volume I and II) - Moon and Mars Exploration Program - Ares Rockets and Orion Spacecraft*, avail: http://www.thebookishblog.com/nasa-s-constellation-program-lessons-learned-volume-i-and-ii.pdf

NASA, *Columbia Accident Investigation Board Report*, Vol. 1 of 6, 2003, ISBN-*0160679044*.

NASA, Saturn-V Flight Manual, 2011, ISBN-1935700707.

National Research Council, Reusable Launch Vehicle: Technology Development and Test Program, 1996, ISBN-978-0309054379.

Orloff. Richard W. and Harland, David M. *Apollo: The Definitive Sourcebook*, Springer Praxis Books/Space Exploration, Springer; 1st ed., 200), ISBN-10: 0387300430.

O'Shaughnessy, Tam *Sally Ride: A Photobiography of America's Pioneering Woman in Space,* 2015, ASIN-B014CS5KSM.

Pelton, Joseph, Marshall, Peter, *License to Orbit: The Future of Commercial Space Travel (Apogee Books Space Series),* 2009, Apogee Books, ISBN-1894959981.

Powell-Willhite, Irene E. *The Voice of Dr. Werner von Braun*, 2007, Apollo Books, ISBN-978-1-894959-64-3.

Presidential Commission (Rogers Commission), *Report on the Space Shuttle Challenger Accident, 1986*, US Government, 5 volumes, Vol. 1 – main report, ISBN-

999739769X.

Rahman, Shamim Worldwide Space Launch Vehicles and Their Mainstage Liquid Rocket Propulsion, 2013, ISBN-978-1289090333.

Rampino, Michael A. Concepts of Operations for a Reusable Launch Vehicle, 2012, ISBN-1249358138.

Rao, PV Manoranjan, Radhakrishnan, P., *A Brief History of Rocketry in ISRO,* Universities Press Private Ltd., 2012, ASIN-B008SC4FM6.

Reichl, David *Project Mercury: America in Space Series,* 1st Edition, Schiffer; 2016, ISBN-0764350692.

Reichl, David *Project Gemini: America in Space Series, 1*st. Edition, Schiffer, 2016, ISBN-0764350706.

Seedhouse, Erik *Spaceports Around the World, A Global Growth Industry* (Springer Briefs in Space Development)1st ed., 2017, Springer, ISBN-3319468456.

Seedhouse, Erik *XCOR, Developing the Next Generation Spaceplane* (Springer Praxis Books), 1st ed. 2016, ISBN-3319468456.

Seedhouse, Erik *SpaceX's Dragon: America's Next Generation Spacecraft,* 2016, Springer, ISBN-3319215140.

Seedhouse, Erik *SpaceX: Making Commercial*

Spaceflight a Reality, 2013, Praxis, ISBN-1461455138.

Seedhouse, Erik *Virgin Galactic: The First Ten Years*, Springer, 2015, ISBN-3319092618.

Seedhouse, Erik *Bigelow Aerospace: Colonizing Space One Module at a Time*, 2015, Springer, ISBN-3319051962.

Shayler David J. *Assembling and Supplying the ISS: The Space Shuttle Fulfills Its Mission,* Springer Praxis Books, 2017, ISBN-3319404415.

Shayler, David J, *Linking the Space Shuttle and Space Stations: Early Docking Technologies from Concept to Implementation*, 2107, Springer ISBN-3319497685.

Shelton, William R. *Man's Conquest of Space*, 1968, National Geographic, ASIN-B000NPONZ0.

Silverberg, Robert *First American Into Space*, Monarch Books, 1961, ASIN-B018M3KN76.

Simpson, Theodore R. *The Space Station, An Idea Whose Time has Come,* IEEE, 1985, ISBN-0879421827.

Sivolella, Davide *The Space Shuttle Program: Technologies and Accomplishments* (Springer Praxis Books), 2017, ISBN-13319549448.

Steven-Boniecki, Dwight *Skylab 1 & 2*, 2015, Apogee

Books, ISBN-1926592271.

Steven-Boniecki, Dwight *Skylab 3, The NASA Mission Reports,* 2016 Apogee Books, ISBN-192659228X.

Stoff, Joshua *Building Moonships, The Grumman Lunar Module*, 2004, Arcadia Press, ISBN-1531620906.

Storms, Dr. Harrison A. and Hallion, Dr. Richard P. *X-15: Reaching for Space* (The X-Plane Series), 2013, ASIN-B00CLIH4Q2.

Swenson, Loyd S. *This New Ocean: A History of Project Mercury,* NASA SP-4201, 1966, ASIN-B0006BP7QA.

Tereshkova, Valentina *Valentina Tereshkova: The First Lady of Space: In Her Own Words,* 2015, ASIN-B017081QG6.

Thompson, Milton O. Armstrong, Neil A. *At the Edge of Space: The X-15 Flight Program,* 2013, Smithsonian Books, ASIN-B00DFIDW9K.

Turkina, Dr. Olwska, Murray, Damon *Soviet Space Dogs,* FUEL Publishing, 2014, ISBN-0956896286.

U.S. GAO *Space Shuttle Accident, NASA's Actions to Address the Presidential Commission Report,* October, 1987, GAO/NSIAD-88-30BR.

U. S. Government, *The Hypersonic Revolution: Case*

Studies in the History of Hypersonic Technology, Volume 1 - From Max Valier to Project PRIME (1924-1967), X-15, X-20A Dyna-Soar, Winged and Lifting Reentry, Rockets, 2015, ASIN-B00T0HC4NU.

U. S. Government, *The Rise and Fall of Dyna-Soar: A History of Air Force Hypersonic R&D, 1944-1963 - Pathfinding Effort to Develop a Trans atmospheric Boost Glider and Spaceplane, Manned Military Space Program,* 2015, ASIN-B00VJK0NNY.

U. S. Government, NASA, *NASA Commercial Crew Human Spaceflight Program for Transport to the International Space Station (ISS): SpaceX Dragon and Boeing CST-100 Contracts, Safety Reviews, History and Update Reports*, 2015, ASIN-B00UCU2OHY.

U. S. Government, DoD, USAF, Center for the Study of National Reconnaissance, *The Dorian Files Revealed: A Compendium of the NRO's Manned Orbiting Laboratory (NRO) Documents, Photo-reconnaissance, Spy in the Sky, Blue Gemini, Air Force Space Station, Dyna-Soar, Apollo Study*, 2017, ISBN-1521163073.

Vance, Ashlee *Elon Musk: Tesla, SpaceX, and the Quest for a Fantastic Future*, 2015, Ecco, ISBN-0062301233.

Van Pelt, Michel *Rocketing Into the Future: The History and Technology of Rocket Planes*, 2012, Springer Praxis, ISBN-1461431999.

von Braun, Wernher; Ordway, Frederick I., III; Durant, Fred *Space Travel: A History: An Update of History of Rocketry & Space Travel*, 1985, ISBN-0061818984.

von Braun, Wernher *Project MARS: A Technical Tale, 2006,* ISBN-0973820330.

Weil, Elizabeth *They All Laughed at Christopher Columbus: An Incurable Dreamer Builds the First Civilian Spaceship*, 2002, Bantam, ISBN-0553108867.

Weiren. Wu *Shenzhou Spacecraft and lunar exploration project* (China's space exploration (English edition), 2008, China Intercontinental Press, ASIN-B00N8BN78U.

Woods, David, Harland, David M. *NASA Gemini 1965-1966, Owners' Workshop Manual*, Haynes, 2015, ISBN-0857334212.

Resources

- www.nasa.gov

- https://history.nasa.gov/ tindex.html#5

- https://www.nasa.gov/exploration/systems/orion/index.html

- https://history.nasa.gov/tindex.html#5

- http://klabs.org/history

- https://www.nasaspaceflight.com

- https://www.nasa.gov/sites/default/files/files/Orion_AZ_book.pdf

- Encyclopedia Astronautica, http://www.astronautix.com/

- current locations of Apollo hardware: https://nssdc.gsfc.nasa.gov/planetary/lunar/apollo loc.html

- Chinese Crewed Space Program https://web.archive.org/web/20070408034953/http://cns.miis.edu/pubs/week/031006.htm

- Soyuz MS-1 http://space.skyrocket.de/doc_sdat/soyuz-ms.htm

- http://www.antonov.com/aircraft/transport-aircraft/an-225-mriya

- Vectors website - http://vc.airvectors.net/idx_sci.html

- https://en.wikipedia.org/wiki/Human_spaceflight

- Wikipedia, various

Glossary

AFB – Air Force base.
ASDS – autonomous spaceport drone ships.
AGC – Apollo Guidance Computer
AIAA – American Institute of Aeronautics and Astronautics.
ALU – arithmetic logic unit
Angkasawan - Malay word for astronaut.
APAS (China) - Androgynous Peripheral Attach System.
Apogee – farthest point in the orbit from the Earth.
ARPA – Advanced Research projects Agency.
ASIN – Amazon Standard Inventory Number
Astrionics – electronics for space flight.
ATOLL - Acceptance Test or Launch Language.
BEO – beyond Earth orbit.
Blooster – balloon-based launch vehicle.
BP – boilerplate. Mechanical model.
C3PO – NASA Commercial Crew and Cargo Program.
CCDev – commercial crew development.
CctCap – Commercial Crew Transportation Capability
COTS – commercial, off the shelf; also commercial orbital transportation services.
CCP- Commercial Crew Program (NASA)
Cpu – central processing unit
CRS – commercial resupply services.
Cyrogenic – pertaining to very low temperatures.
DoD – (U. S.) Department of Defense.
DTM – dynamic test model, for structural tests.
ECLSS – Environmental Control & Life Support system.
EELV – Evolved Expendable Launch Vehicle, 1996

Program.
Ephemeris – position information data set for orbiting bodies, 6 parameters plus time.
ETR – Eastern Test Range, Cape Canaveral, FL
EVA – extra-vehicular activity
FAA – Federal Aviation Administration
Gimbal – pivoted support, allowing rotation about 1 axis.
GLS – (lunar) gateway logistics services.
Gpm – gallons per minute.
GSFC – NASA Goddard Space Flight Center, Greenbelt, MD.
Gyro – device to measure angular rate.
H1 – Rocketdyne engine, used on Saturn-I first stage.
HST – Hubble Space Telescope
Hypersonic – flight at Mach 5 and above (Mach is the speed of sound, at the relevant altitude).
ICBM – Intercontinental Ballistic Missile.
IBM – International Business Machines Company.
IDIQ – Indefinite delivery/indefinite quanty contract.
ISBN – international standard book number.
ISP – specific impulse. Measure of efficiency of rocket engine. Units of seconds.
ISRO – Indian Space Research Organization.
ITAR – (U.S.) International Traffic in Arms Regulations
IU – Instrument Unit, Saturn launch vehicle.
IUS – Interim Upper Stage,.JPL – Jet Propulsion Laboratory, Pasadena, CA.
JSC – Johnson Space Center, Houston, Texas.
Jupiter – ICBM, 3-stage. Developed by von Braun Team.
Kámán line – international definition of the beginning of Space. 100 Km above the surface.

Kbps – kilo (10^3) bits per second.
Khz – kilohertz, one thousand cycles per second.
Kev – kilo electron volts, measure of energy of a particle.
KSC – NASA Kennedy Space Center, launch site, Florida.
Hypersonic – 5 times the speed of sounds, and beyond.
IUS – Interim Upper Stage (for STS).
LAS – launch abort system.
Lb – pound weight.
Lbf – pounds, force.
LC-37 – Launch Complex – 37 at KSC.
LEM – lunar excursion module.
LEO – low Earth orbit.
LES – Apollo Launch Escape System.
LH2 – liquid hydrogen.
LOX – liquid oxygen, boils at -297 F.
LVDA – Launch Vehicle Data Adapter.
LVDC – Launch Vehicle Digital Computer.
Mach – unit of speed at sound, named after E. Mach, aerodynamics pioneer.
Mev – million electron volts, measure of energy of a particle.
MINITRACK – "Minimum Trackable Satellite " U. S. satellite tracking network, 1957.
MIT – Massachusetts Institute of Technology
MOL – Manned Orbiting Lab. Gemini-era space station by USAF.
MSC – Manned Space Center, Houston, TX. Renamed JSC.
MSFC – NASA Marshall Space Flight Center, Huntsville, AL.

m/s – meters per second.
NACA – National Advisory Committee for Aeronautics, predecessor of NASA.
NASA – National Aeronautics and Space Administration.
NASCOM – NASA Communications Network. Worldwide, operated by GSFC.
NEO – near Earth orbit.
NOR – negative or logic
NORAD – North American Air Defense.
NRL – Naval Research Lab, Washington, DC.
NRO – (U. S.) National Reconnaissance Officve
NSSL – (U. S.) National Security Space launch.
NTIS – National Technical Information Service, (www.ntis.gov).
OAMS – orbit attitude and maneuvering unit.
Ogive – a pointed arch structure.
ORU – orbital replacement unit.
PAM – pulse amplitude modulation.
Pc – personal computer
PCM – pulse code modulation.
Perigee – closest point in the orbit from the Earth.
PGNCS – Primary Guidance, Navigation, and Control system for Apollo.
PGSC – portable general support computer (ISS)
POGO – longitudinal oscillation in liquid-fueled rocket motors that can lead to failure.
Pregnant Guppy – large cargo aircraft operated by Aero Spacelines 1963-1979.
RAM – Random Access Memory – generally, read-write.
RCS – reaction control system

R&D – research & development.
Redstone – Army missile developed by the von Braun team. Used for Mercury crewed flights.
Redstone Arsenal – Army R&D facility in Huntsville, AL. Later became NASA MSFC.
ROM – Read-Only Memory, used for storage of instructions and fixed data.
RP-1 – rocket propellant-one, highly refined kerosene.
RTLS – return to launch site.
SA – x – Saturn-Apollo – flight x.
SAO – Smithsonian Astrophysical Observatory.
SI – System International – the metric system.
S-IC – first stage of the Saturn V
S-II – second stage of the Saturn V
S-IVB – third stage of the Saturn V
S-IV – second stage of Saturn 1 rocket.
Shenzhou – Divine Ark, in Chinese.
SLS – Space Launch System (NASA)
Soyuz – Union, in Russian
STADAN – Space Tracking and Data Acquisition Network.
STS – Space transportation system – shuttle.
TDRS – Tracking andData Relay Satellites.
Titan – ICBM and NASA/USAF launch vehicle.
TKS – Russian transport supply spacecraft.
TM – Technical Manual.
Ullage – residual fuel or oxidizer in a tank after engine burn is complete.
V-2 – German World War-II missile developed by the von Braun Team.
Vdc – volts, direct current.

Voskhod – Sunrise, in Russian
Vostok – East, in Russian
WSMR – White Sands Missile Range, New Mexico.
Zombie-Sat – dead satellite in orbit.

If you enjoyed this book, you might also be interested in some of these.

Stakem, Patrick H. *16-bit Microprocessors, History and Architecture*, 2013 PRRB Publishing, ISBN-1520210922.

Stakem, Patrick H. *4- and 8-bit Microprocessors, Architecture and History*, 2013, PRRB Publishing, ISBN-152021572X,

Stakem, Patrick H. *Apollo's Computers*, 2014, PRRB Publishing, ISBN-1520215800.

Stakem, Patrick H. *The Architecture and Applications of the ARM Microprocessors,* 2013, PRRB Publishing, ISBN-1520215843.

Stakem, Patrick H. *Earth Rovers: for Exploration and Environmental Monitoring,* 2014, PRRB Publishing, ISBN-152021586X.

Stakem, Patrick H. *Embedded Computer Systems, Volume 1, Introduction and Architecture*, 2013, PRRB Publishing, ISBN-1520215959.

Stakem, Patrick H. *The History of Spacecraft Computers from the V-2 to the Space Station*, 2013, PRRB Publishing, ISBN-1520216181.

Stakem, Patrick H. *Floating Point Computation*, 2013, PRRB Publishing, ISBN-152021619X.

Stakem, Patrick H. *Architecture of Massively Parallel Microprocessor Systems*, 2011, PRRB Publishing, ISBN-1520250061.

Stakem, Patrick H. *Multicore Computer Architecture*, 2014, PRRB Publishing, ISBN-1520241372.

Stakem, Patrick H. *Personal Robots*, 2014, PRRB Publishing, ISBN-1520216254.

Stakem, Patrick H. *RISC Microprocessors, History and Overview,* 2013, PRRB Publishing, ISBN-1520216289.

Stakem, Patrick H. *Robots and Telerobots in Space Application*s, 2011, PRRB Publishing, ISBN-1520210361.

Stakem, Patrick H. *The Saturn Rocket and the Pegasus Missions, 1965,* 2013, PRRB Publishing, ISBN-1520209916.

Stakem, Patrick H. *Visiting the NASA Centers, and Locations of Historic Rockets & Spacecraft,* 2017, PRRB Publishing, ISBN-1549651205.

Stakem, Patrick H. *Microprocessors in Space*, 2011, PRRB Publishing, ISBN-1520216343.

Stakem, Patrick H. Computer *Virtualization and the Cloud*, 2013, PRRB Publishing, ISBN-152021636X.

Stakem, Patrick H. *What's the Worst That Could Happen? Bad Assumptions, Ignorance, Failures and Screw-ups in Engineering Projects, 2014,* PRRB Publishing, ISBN-1520207166.

Stakem, Patrick H. *Computer Architecture & Programming of the Intel x86 Family, 2013,* PRRB Publishing, ISBN-1520263724.

Stakem, Patrick H. *The Hardware and Software Architecture of the Transputer,* 2011,PRRB Publishing, ISBN-152020681X.

Stakem, Patrick H. *Mainframes, Computing on Big Iron*, 2015, PRRB Publishing, ISBN- 1520216459.

Stakem, Patrick H. *Spacecraft Control Centers*, 2015, PRRB Publishing, ISBN-1520200617.

Stakem, Patrick H. *Embedded in Space,* 2015, PRRB Publishing, ISBN-1520215916.

Stakem, Patrick H. *A Practitioner's Guide to RISC Microprocessor Architecture*, Wiley-Interscience, 1996, ISBN-0471130184.

Stakem, Patrick H. *Cubesat Engineering*, PRRB Publishing, 2017, ISBN-1520754019.

Stakem, Patrick H. *Cubesat Operations*, PRRB Publishing, 2017, ISBN-152076717X.

Stakem, Patrick H. *Interplanetary Cubesats*, PRRB Publishing, 2017, ISBN-1520766173.

Stakem, Patrick H. Cubesat Constellations, Clusters, and Swarms, Stakem, PRRB Publishing, 2017, ISBN-1520767544.

Stakem, Patrick H. *Graphics Processing Units, an overview*, 2017, PRRB Publishing, ISBN-1520879695.

Stakem, Patrick H. *Intel Embedded and the Arduino-101, 2017,* PRRB Publishing, ISBN-1520879296.

Stakem, Patrick H. *Orbital Debris, the problem and the mitigation*, 2018, PRRB Publishing, ISBN-*1980466483*.

Stakem, Patrick H. *Manufacturing in Space*, 2018, PRRB Publishing, ISBN-1977076041.

Stakem, Patrick H. *NASA's Ships and Planes*, 2018, PRRB Publishing, ISBN-1977076823.

Stakem, Patrick H. *Space Tourism*, 2018, PRRB Publishing, ISBN-1977073506.

Stakem, Patrick H. *STEM – Data Storage and Communications*, 2018, PRRB Publishing, ISBN-

1977073115.

Stakem, Patrick H. *In-Space Robotic Repair and Servicing*, 2018, PRRB Publishing, ISBN-1980478236.

Stakem, Patrick H. *Introducing Weather in the pre-K to 12 Curricula, A Resource Guide for Educators*, 2017, PRRB Publishing, ISBN-1980638241.

Stakem, Patrick H. *Introducing Astronomy in the pre-K to 12 Curricula, A Resource Guide for Educators*, 2017, PRRB Publishing, ISBN-198104065X.
Also available in a Brazilian Portuguese edition, ISBN-1983106127.

Stakem, Patrick H. *Deep Space Gateways, the Moon and Beyond*, 2017, PRRB Publishing, ISBN-1973465701.

Stakem, Patrick H. *Exploration of the Gas Giants, Space Missions to Jupiter, Saturn, Uranus, and Neptune*, PRRB Publishing, 2018, ISBN-9781717814500.

Stakem, Patrick H. *Crewed Spacecraft*, 2017, PRRB Publishing, ISBN-1549992406.

Stakem, Patrick H. *Rocketplanes to Space*, 2017, PRRB Publishing, ISBN-1549992589.

Stakem, Patrick H. *Crewed Space Stations,* 2017, PRRB Publishing, ISBN-1549992228.

Stakem, Patrick H. *Enviro-bots for STEM: Using Robotics in the pre-K to 12 Curricula, A Resource Guide for Educators,* 2017, PRRB Publishing, ISBN-1549656619.

Stakem, Patrick H. *STEM-Sat, Using Cubesats in the pre-K to 12 Curricula, A Resource Guide for Educators*, 2017, ISBN-1549656376.

Stakem, Patrick H. *Lunar Orbital Platform-Gateway*, 2018, PRRB Publishing, ISBN-1980498628.

Stakem, Patrick H. *Embedded GPU's*, 2018, PRRB Publishing, ISBN- 1980476497.

Stakem, Patrick H. *Mobile Cloud Robotics*, 2018, PRRB Publishing, ISBN- 1980488088.

Stakem, Patrick H. *Extreme Environment Embedded Systems,* 2017, PRRB Publishing, ISBN-1520215967.

Stakem, Patrick H. *What's the Worst, Volume-2*, 2018, ISBN-1981005579.

Stakem, Patrick H., *Spaceports*, 2018, ISBN-1981022287.

Stakem, Patrick H., *Space Launch Vehicles*, 2018, ISBN-1983071773.

Stakem, Patrick H. *Mars*, 2018, ISBN-1983116902.

Stakem, Patrick H. *X-86, 40th Anniversary ed*, 2018, ISBN-1983189405.

Stakem, Patrick H. *Lunar Orbital Platform-Gateway*, 2018, PRRB Publishing, ISBN-1980498628.

Stakem, Patrick H. *Space Weather*, 2018, ISBN-1723904023.

Stakem, Patrick H. *STEM-Engineering Process*, 2017, ISBN-1983196517.

Stakem, Patrick H. *Space Telescopes,* 2018, PRRB Publishing, ISBN-1728728568.

Stakem, Patrick H. *Exoplanets*, 2018, PRRB Publishing, ISBN-9781731385055.

Stakem, Patrick H. *Planetary Defense*, 2018, PRRB Publishing, ISBN-9781731001207.

Patrick H. Stakem *Exploration of the Asteroid Belt*, 2018, PRRB Publishing, ISBN-1731049846.

Patrick H. Stakem *Terraforming*, 2018, PRRB Publishing, ISBN-1790308100.

Patrick H. Stakem, *Martian Railroad,* 2019, PRRB Publishing, ISBN-1794488243.

Patrick H. Stakem, *Exoplanets,* 2019, PRRB Publishing, ISBN-1731385056.

Patrick H. Stakem, *Exploiting the Moon,* 2019, PRRB Publishing, ISBN-1091057850.

Patrick H. Stakem, *RISC-V, an Open Source Solution for Space Flight Computers,* 2019, PRRB Publishing, ISBN-1796434388.

Patrick H. Stakem, *Arm in Space*, 2019, PRRB Publishing, ISBN-9781099789137.

Patrick H. Stakem, *Extraterrestrial Life*, 2019, PRRB Publishing, ISBN-978-1072072188.

Patrick H. Stakem, *Space Command*, 2019, PRRB Publishing, ISBN-978-1693005398.

CubeRovers, A Synergy of Technologys, 2020, PRRB Publishing, ISBN-979-8651773138.

Robotic Exploration of the Icy moons of the Gas Giants. 2020, PRRB Publishing, ISBN- 979-8621431006

Hacking Cubesats, 2020, PRRB Publishing, ISBN-979-8623458964.

History & Future of Cubesats, PRRB Publishing, ISBN-979-8649179386.

Hacking Cubesats, Cybersecurity in Space, 2020, PRRB Publishing, ISBN-979-8623458964.

Powerships, Powerbarges, Floating Wind Farms: electricity when and where you need it, 2021, PRRB Publishing, ISBN-979-8716199477.

Hospital Ships, Trains, and Aircraft, 2020, PRRB Publishing, ISBN-979-8642944349.

2020/2021 Releases

CubeRovers, a Synergy of Technologys, 2020, ISBN-979-8651773138

Exploration of Lunar & Martian Lava Tubes by Cube-X, ISBN-979-8621435325.

Robotic Exploration of the Icy moons of the Gas Giants, ISBN- 979-8621431006.

History & Future of Cubesats, ISBN-978-1986536356.

Robotic Exploration of the Icy Moons of the Ice Giants, by Swarms of Cubesats, ISBN-979-8621431006.

Swarm Robotics, ISBN-979-8534505948.

Introduction to Electric Power Systems, ISBN-979-8519208727.

Centros de Control: Operaciones en Satélites del Estándar CubeSat (Spanish Edition), 2021, ISBN-979-8510113068.

Exploration of Venus, 2022, ISBN-979-8484416110.

Patrick H. Stakem, *The Search for Extraterrestial Life,* 2019, PRRB Publishing, ISBN-1072072181.

The Artemis Missions, Return to the Moon, and on to Mars, 2021, ISBN-979-8490532361.

James Webb Space Telescope. A New Era in Astronomy, 2021, ISBN-979-8773857969.

www.ingramcontent.com/pod-product-compliance
Lightning Source LLC
Chambersburg PA
CBHW020928180526
45163CB00007B/2932